Prabhakar Shukla

Performance Evaluation of Conservation Programmes for Lakes of the Nainital Region

GRIN Verlag

Bibliografische Information der Deutschen Nationalbibliothek:

Die Deutsche Bibliothek verzeichnet diese Publikation in der Deutschen Nationalbibliografie; detaillierte bibliografische Daten sind im Internet über http://dnb.d-nb.de/ abrufbar.

Imprint:

Druck und Bindung: Books on Demand GmbH, Norderstedt Germany
ISBN: 978-3-656-68608-8

This book at GRIN:

http://www.grin.com/en/e-book/275593/performance-evaluation-of-conservation-programmes-for-lakes-of-the-nainital

PERFORMANCE EVALUATION OF CONSERVATION PROGRAMMES FOR LAKES OF NAINITAL REGION

Hitesh Dalakoti
Alternate Hydro Energy Center
Indian Institute of Technology
Roorkee,India

Sunil Kumar Singal
Alternate Hydro Energy Center
Indian Institute of Technology
Roorkee,India

Abstract: *The lakes of Nainital region are ecologically fragile, and due to rapid urbanization in the catchment, the lakes experience an increase in nutrient loading and rapid deterioration of water quality. The present work was done to assess the effectiveness of the conservation works done in the lakes of Nainital, Bhimtal, Sattal and Naukuchiatal. The various aspects that were kept in mind during the study were the water quality assessment, solid waste management practices, shoreline development, sewage works, public participation. Various physico- chemical parameters were determined during the water quality assessment like pH, temperature, DO, BOD,COD, turbidity, TDS, and metals namely Copper, Iron, Nitrate, and Chromium. These parameters were compared with the trends of the physico-chemical parameters in the recent years. A water quality index was prepared on the basis of these parameters to make an evaluation of the health of the lakes. Noteworthy improvement in the dissolved oxygen concentrations of the lakes were an evidence of the fact that significant work has been done regarding the conservation of the lakes. All the shortcomings and success regarding the conservation works done in the area have been studied and a performance evaluation chart was prepared on this basis. The results of the study suggested that effective works have been done in the area but this needs to be continued in the future too.*

Key words: *Water quality, Nainital lake, Bhimtal lake, Sattal lake, Naukuchiatal lake, Performance evaluation*

INTRODUCTION

A lake is just an added constituent of Earth's surface water. A lake is the place where surface-water runoff gets collected in a low mark, in relation to the neighboring countryside. It's not that the stream that structures lakes get fascinated, but that the stream inflowing a lake approaches faster than it can get away, either via seeping away in a river, discharge into the land, or by evaporation. Nearly every lake contains fresh water, but few, mostly those where stream cannot getaway through river, are saline lakes. Almost all the lakes hold a lot of water organisms, but not every aquatic life. The conservation programmes are being implemented on these lakes time to time. Such conservation programmes and their monitoring are necessary. The evaluation of various conservation measures are necessary to check quality of seriousness of the work done. Moreover, the general public is also conscious about conservation of the lake resources because they are affected most. Some NGOs and concerned people are making every effort for conservation. Considering alone, an attempt has been made to review on the performance evaluation of the conservation work going on/ carried out on the lakes in Nainital area.

STUDY AREA

Nainital is a glittering jewel in the Himalyan necklace, blessed with scenic natural spledour and varied natural resources . Dotted with lakes , Nainital has earned the epithet of ' **Lake District** ' of India . The most prominent of the lakes is Naini lake ringed by hills . Nainital has a varied topography . Some of the important places in the district are Nainital , Haldwani , Kaladhungi , Ramnagar , Bhowali , Ramgarh , Mukteshwar , Bhimtal , Sattal and Naukuchiatal . Nainital's unending expense of scenic beauty is nothing short of a romance with awe-inspiring and pristine Mother nature.Nainital is a trendy hill location in Uttarakhand state and center of operations of Nainital district , located at an elevation of 2,084 metres (6,837 ft) beyond MSL. Bhimtal is a municipality and a nagar panchayat in Nainital district in Uttarakhand, located at an elevation of 1370 meters beyond sea level and is a 22 kilometres commencing Nainital. Bhimtal is a municipality and a nagar panchayat in Nainital district in Uttarakhand, located at an elevation of 1370 meters beyond sea level and is a 22 kilometres commencing Nainital. Naukuchiatal also called lake of nine corners' is a self-effacing hilly location in Nainital district ,Uttarakhand. The lake is 175 feet in depth and is located at 1220 mts beyond mean sea level. Sattal / Sat Tal (English name "seven lakes") is an consistent assembly of seven freshwater lake located in Lower Himalayan Range close to Bhimtal,

Nainital district , Uttarakhand.The location of the lakes has been shown in **figure 1** and the morphometry and the longitudinal and latitudinal locations of the lakes have been given in **table 1.**

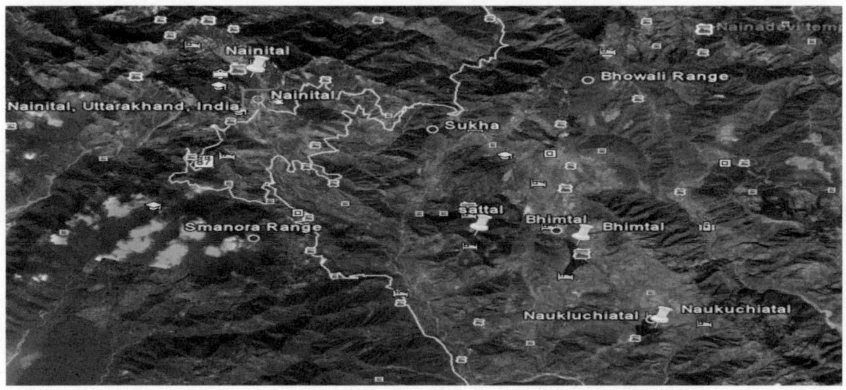

Fig. 1: Location of Lakes **(Source: Google Earth Image)**

Table 1: Location and Morphometry of Lakes of Nainital

S. No.	Parameters	Nainital	Bhimtal	Naukuchiatal	Sattal
1	Altitude(m)	1937	1331	1220	1286
2	Latitude	29°24'N	29°20'N	29°19'N	29°19'N
3	Length(m)	1432	1974	950.9	No data
4	Width(m)	423	457	692	No data
5	Max.depth (meter)	27.3	25.8	40.8	20.0
6	Mean depth (meter)	16.2	11.5	No data	No data
7	exterior area(ha)	48.2	72.3	44.5	No data
8	Catchment area(ha)	397.7	975.4	No data	332.0
9	Shoreline(m)	3630	4023	3560	No data

MATERIALS AND METHODS

Samples were collected as per (APHA,2005) and sample testing was done at the laboratory in the AHEC, IITR. In brief each sample was considered to be a composite representation of the whole lake taken from various sampling points selected according to their relevancy. Samples were collected from surface of the lakes using standard water sampler, which were then kept in large air-tight plastic air containers at 4^0C and were transported within 10 hours of their collection. Sample for metal analysis and for physico- chemical analysis were kept in different bottles with specific preservative(APHA,2005). Temperature was measured at the site with the help of thermometer. The whole study is divided into two stages. In the first stage sample were collected in the month of November and in the second stage sample are to be collected in the month of April. This was done to determine the water quality and the environmental conditions of the lake presently in comparison to the previous steps towards the lake conservation. The concentrations of nitrate, chromium, iron were measured by Hach make spectrophotometer. These equipments are preprogrammed and follow the methods of APHA (2005). The reagents for these experiments are supplied by the manufacturers themselves and are calibrated with equipment to display direct reading/ concentrations. Biochemical Oxygen Demand (BOD) was determined following APHA (2005). Water samples were incubated at 20^0 C for 5 days in a BOD incubator. Initial and final readings of DO were measured by a laboratory DO meter, attached with a BOD probe

WATER QUALITY SAMPLING

Water samples were collected from the study area. The sampling locations were selected on the basis of their proneness to pollution, approachability etc. On the basis of above criteria, sampling points were points were selected which is shown in **table 2.** To study the various effects on the lakes, water samples were collected from various points of the lakes were taken and analyzed. Samples need to be collected two times, once in November,2013 and secondly in April, 2014.

Table 2: Sampling Locations

S.No.	Sample	Description
1	N1	NEAR BOAT HOUSE CLUB, MALLITAL
2	N2	PHANSI GADHERA, TALLITAL
3	N3	NEAR PASHAN DEVI MANDIR, THANDI SADAK
4	NA1	CHANOUTI VILLAGE, UPPER SIDE
5	NA2	CHANOUTI VILLAGE, UPPER SIDE
6	NA3	NEAR BOAT STAND
7	B1	MALLITAL TIKONIA
8	B2	KAINCHULI DEVI MANDIR
9	B3	DAAT
10	S1	NEAR SATTAL CHRISTAIN ESTATE
11	S2	NEAR SURIYAGAON
12	S3	MID LAKE

RESULTS AND DISCUSSION

Sampling was done through various experimental methods to check the various physico-chemical parameters along with the metal analysis. The results of the sampling have been tabulated from **table 3** to **table 6** for two seasons.

Table 3: Comparison of physico-chemical and metal analysis of water samples taken from Nainital lake done at 3 sampling points (April, 2014 and November, 2013)

S.No.	Parameters	Near Boat House Club,Mallital		Phansi Gadhera, Tallital		Near Pashan Devi Mandir, Thandi Sadak		Normal Desirable value
		April, 2014	Nov, 2013	April, 2014	Nov, 2013	April, 2014	Nov, 2013	
1	Temp. (^{0}C)	15.5	13.5	15.7	13.0	14.6	12.0	-
2	pH	7.6	7.9	7.3	7.6	7.1	7.4	6.5-8.5
3	DO (mg/l)	8.7	8.9	7.8	8.5	8.2	8.7	5-12
4	BOD (mg/l)	3.7	3.1	3.8	3.5	2.7	2.5	<6
5	COD (mg/l)	5.8	6.0	5.5	5.9	6.1	6.0	
6	Turbidity (NTU)	8.6	8.25	7.3	7.8	7.6	7.6	
7	TDS (mg/l)	230	225	226	224	234	230	<500
8	Cu (mg/l)	1.09	1.01	0.89	0.96	1.03	1.09	<5
9	Fe (mg/l)	0.65	0.67	0.15	0.14	0.44	0.46	<1.5
10	Nitrate (mg/l)	0.26	0.29	0.33	0.37	0.37	0.35	<0.3
11	Cr (mg/l)	0.16	0.15	0.08	0.06	0.12	0.16	

Table 4: Comparison of physicochemical and metal analysis of water samples taken from Bhimtal lake done at three sampling points (April, 2014 and and November, 2013)

S.No.	Parameters	Mallital Tikonia		Kainchuli Devi Mandir		Daat		Normal Desirable value
		April, 2014	Nov, 2013	April, 2014	Nov, 2013	April, 2014	Nov, 2013	
1	Temp. (^{0}C)	17.8	15.5	16.5	16.0	17.2	17.0	-
2	pH	7.5	7.8	7.9	7.6	8.1	7.5	6.5-8.5
3	DO (mg/l)	7.9	7.8	7.1	6.9	7.9	7.9	5-12
4	BOD (mg/l)	2.6	2.6	2.7	2.3	2.4	2.7	<6
5	COD (mg/l)	5.9	5.7	5.8	5.6	5.4	5.5	
6	Turbidity (NTU)	4.6	4.3	4.1	3.9	3.6	3.7	
7	TDS (mg/l)	120	130	128	125	122	120	<500
8	Cu (mg/l)	0.038	0.045	0.045	0.036	0.052	0.053	<5
9	Fe (mg/l)	0.033	0.032	0.032	0.026	0.025	0.024	<1.5
10	Nitrate (mg/l)	0.56	0.57	0.45	0.48	0.6	0.47	<0.3
11	Cr (mg/l)	0.17	0.18	0.16	0.14	0.13	0.22	

Table 5: Comparison of physicochemical and metal analysis of water samples taken from Naukuchiatal lake done at three sampling points (April , 2014 and November, 2013)

S.No.	Parameters	Chanouti village,Upper side		Chanouti Village,middle		Near boat stand		Normal Desirable value
		April, 2014	Nov, 2013	April, 2014	Nov, 2013	April, 2014	Nov, 2013	
1	Temp. (^{0}C)	17.2	16.7	15.0	15.4	15.8	15.8	-
2	pH	8.2	8.5	8.3	8.4	8.1	8.2	6.5-8.5
3	DO (mg/l)	5.6	7.9	6.1	7.1	5.4	7.9	5-12
4	BOD (mg/l)	7.4	7.9	8.3	8.7	6.1	6.1	<6
5	COD (mg/l)	5.5	5.9	6.2	6.3	6.1	6.5	
6	Turbidity (NTU)	120	123	112	110	114	112	<500
7	TDS (mg/l)	0.033	0.034	0.026	0.032	0.028	0.025	<5
8	Cu (mg/l)	0.015	0.019	0.017	0.013	0.016	0.011	<1.5
9	Fe (mg/l)	0.34	0.54	0.41	0.43	0.40	0.44	<0.30
10	Nitrate (mg/l)	0.13	0.10	0.20	0.21	0.15	0.15	
11	Cr (mg/l)	8.5	8.8	8.4	9.1	9.2	8.6	

Table 6: Comparison of physicochemical and metal analysis of water samples taken from Sattal lake done at three sampling points (April, 2014 and November, 2013)

S.No.	Parameters	Near Sattal Christian Estate		Near Suriyagaon		Mid lake		Normal Desirable value
		April, 2014	Nov, 2013	April, 2014	Nov, 2013	April, 2014	Nov, 2013	
1	Temp. (^0C)	18.5	18.7	18.8	18.6	19.1	16.7	-
2	pH	8.1	8.3	7.7	7.9	8.4	8.1	6.5-8.5
3	DO (mg/l)	7.4	6.1	7.1	5.8	6.9	5.9	5-12
4	BOD (mg/l)	6.2	6.4	6.2	5.5	5.6	6.3	<6
5	COD (mg/l)	7.4	7.9	7.2	7.1	7.3	7.5	
6	Turbidity (NTU)	7.68	7.65	7.7	7.25	8.1	8.1	
7	TDS (mg/l)	221	223	217	220	215	215	<500
8	Cu (mg/l)	0.018	0.017	0.019	0.016	0.012	0.013	<5
9	Fe (mg/l)	0.028	0.025	0.026	0.026	0.020	0.021	<1.5
10	Nitrate (mg/l)	0.32	0.30	0.28	0.28	0.34	0.31	<0.30
11	Cr (mg/l)	0.13	0.11	0.19	0.18	0.23	0.24	

COMPARATIVE GRAPHS FOR SAMPLING RESULTS IN FOUR LAKES

The following graphs represent the comparative analysis of various parameters at different locations showing the permissible ranges

(A) Nainital Lake

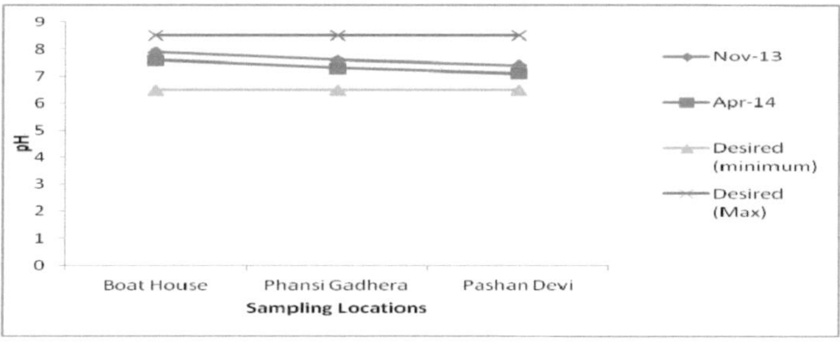

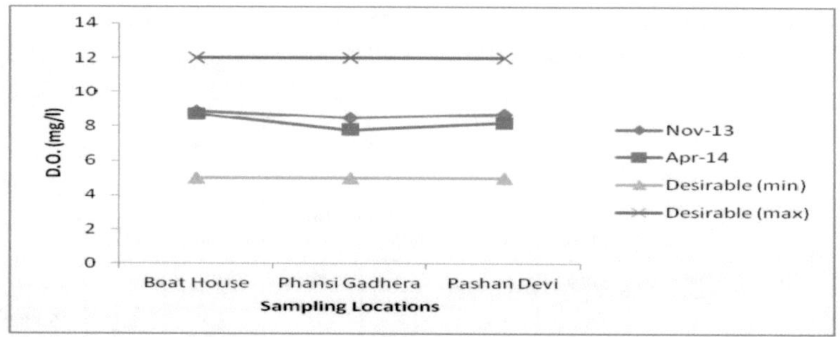
D.O. (mg/l)
14
12
10
8
6
4
2
0
Boat House
Phansi Gadhera
Pashan Devi
Sampling Locations
Nov-13
Apr-14
Desirable (min)
Desirable (max)

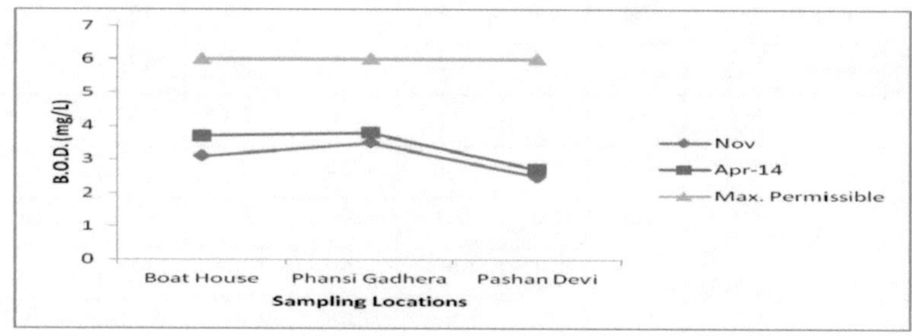
B.O.D. (mg/L)
7
6
5
4
3
2
1
0
Boat House
Phansi Gadhera
Pashan Devi
Sampling Locations
Nov
Apr-14
Max. Permissible

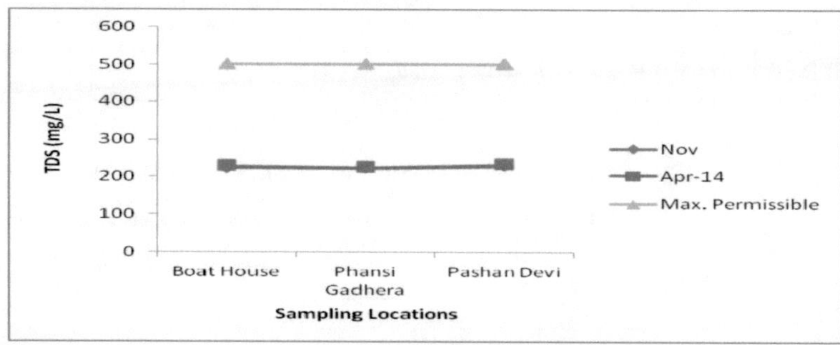
TDS (mg/L)
600
500
400
300
200
100
0
Boat House
Phansi Gadhera
Pashan Devi
Sampling Locations
Nov
Apr-14
Max. Permissible

B. Bhimtal Lake

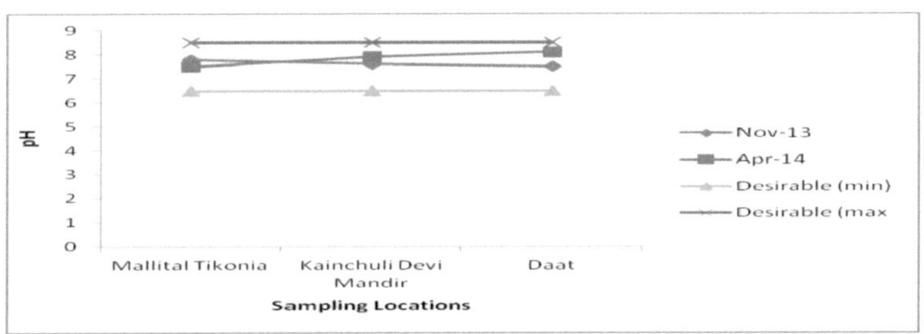

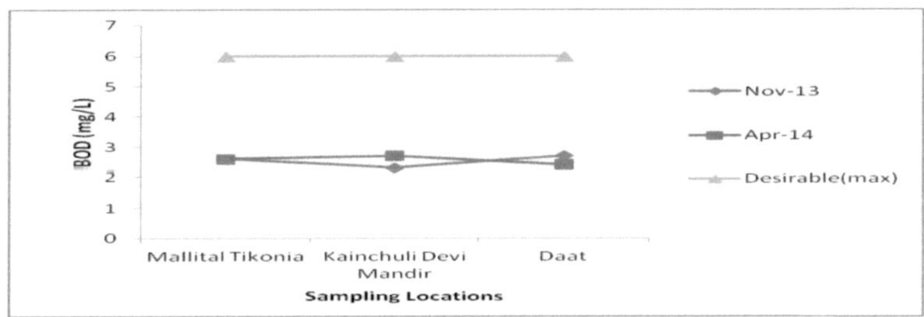

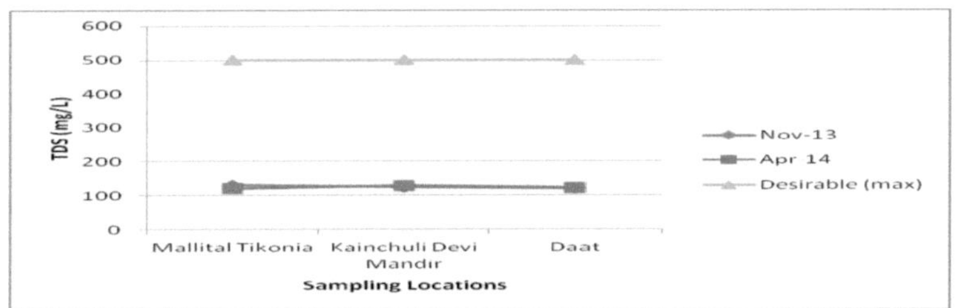

C. Naukuchiatal Lake

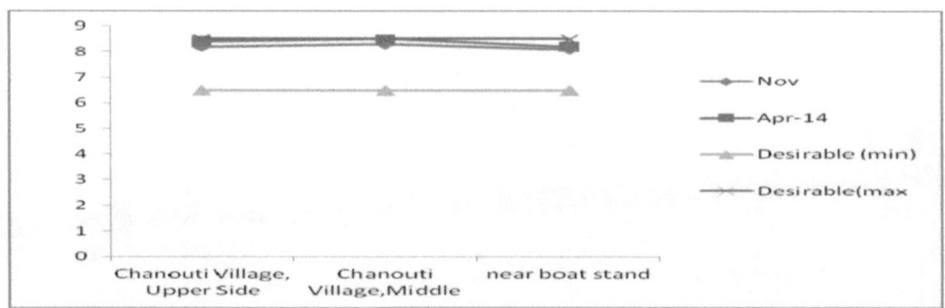

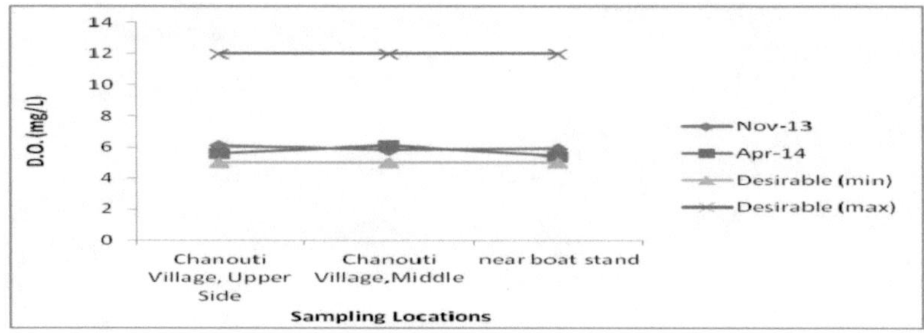

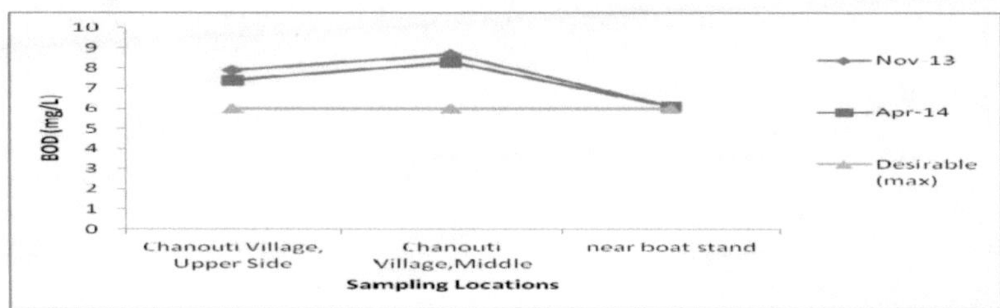

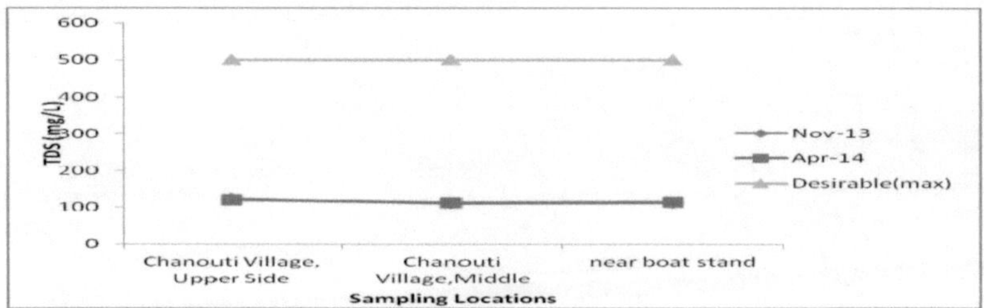

D. Sat Tal Lake

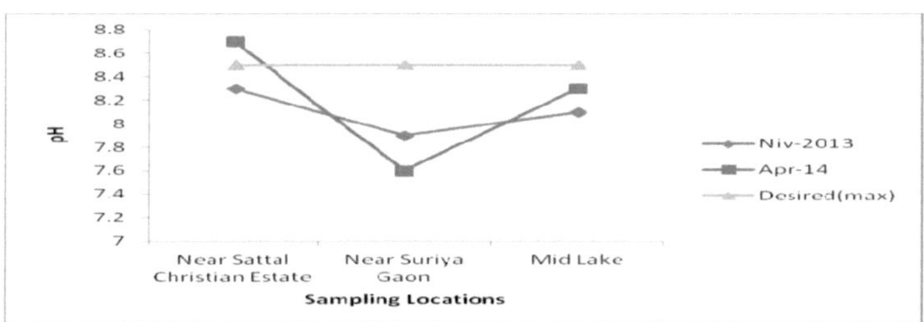

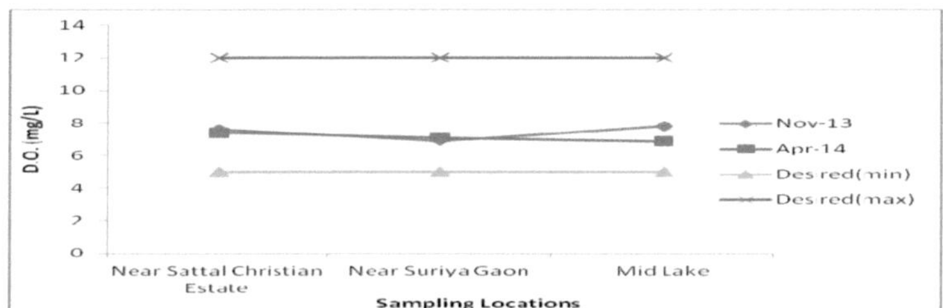

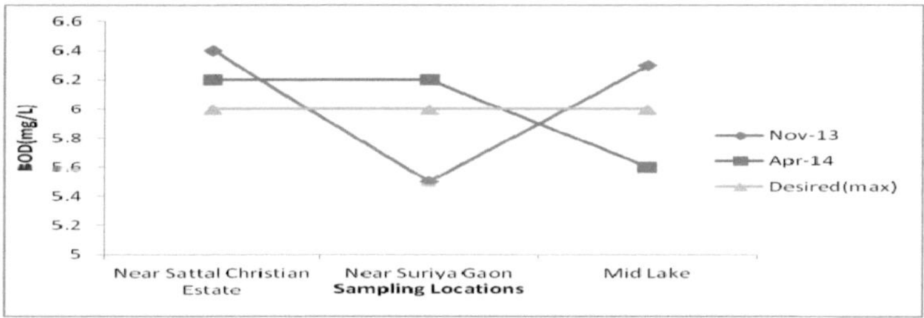

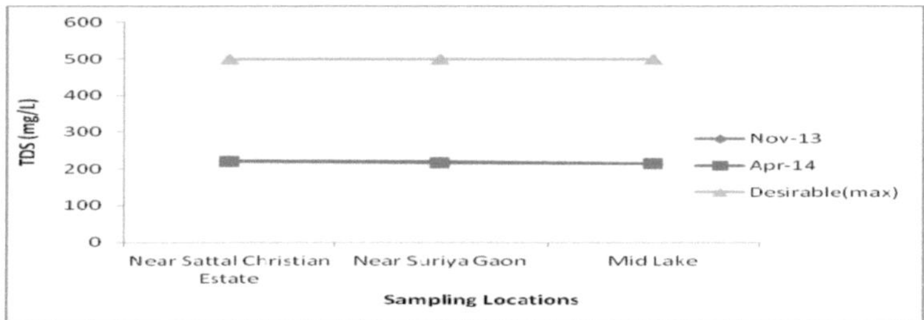

WATER QUALITY INDEX

The water quality index was calculated using 4 important physico- chemical properties namely dissolved oxygen, biochemical oxygen demand, pH and the total solids concentration. These results have been shown in **table 7** and **table 8** for November, 2013 and April, 2014 respectively. The following formula has been used for calculating the water quality index:

$$\sum_{1}^{4}\frac{1}{Xi} = \frac{1}{Xi(pH)} + \frac{1}{Xi(DO)} + \frac{1}{Xi(TDS)} + \frac{1}{Xi(BOD)}$$

Weightage factor, $W_i = k/X_i$, X_i= the minimum permissible limit as recommended by the ICMR

$$k = 1/\textstyle\sum_{1}^{4} 1/Xi.$$

The numerical value and the weighting factor having relative significance are multiplied. Total of the resulting values is done to get an overall WQI (WQI= W_i* Xr).

Table 7 :Water Quality Index of Study Area, November-2013			
Sr. No.	**Sampling Site**	**WQI**	**Quality**
1	**Nainital Lake**		
1.1	Near Boat House Club, Mallital	51	Medium
1.2	Phansi Gadhera	51	Medium
1.3	Near Pashan Devi Mandir	53	Medium
2	**Bhimtal Lake**		
2.1	Mallital Tikonia	53	Medium
2.2	Kainchuli Devi Mandir	54	Medium
2.3	Daat	54	Medium
3	**Naukauchiatal**		
3.1	Chanouti Village,Upper side	46	Bad
3.2	Chanouti Village,Lower Middle side	44	Bad
3.3	Near Boat Stand	47	Bad
4	**Sattal**		
4.1	Near Sattal Christain Estate	44	Bad
4.2	Near Suriya Gaon	47	Bad
4.3	Mid Lake	45	Bad

Table 8: Water Quality Index of Study Area, April-2014			
Sr. No.	**Sampling Site**	**WQI**	**Quality**
1	**Nainital Lake**		
1.1	Near Boat House Club, Mallital	49	Bad
1.2	Phansi Gadhera	50	Bad
1.3	Near Pashan Devi Mandir	52	Medium
2	**Bhimtal Lake**		
2.1	Mallital Tikonia	53	Medium
2.2	Kainchuli Devi Mandir	56	Medium
2.3	Daat	55	Medium
3	**Naukauchiatal**		
3.1	Chanouti Village,Upper side	46	Bad
3.2	Chanouti Village,Lower Middle side	45	Bad
3.3	Near Boat Stand	47	Bad
4	**Sattal**		
4.1	Near Sattal Christain Estate	44	Bad
4.2	Near Suriya Gaon	47	Bad
4.3	Mid Lake	46	Bad

RESULTS AND DISCUSSIONS

An acceptable level of BOD, COD and other parameters suggest that the restoration programmes being implemented in the lakes have given fruitful results. The results of the analysis are discussed below:

1. BOD and COD : Organic pollution is indicated by the BOD value in the aquatic systems, which affect the lake water quality and biodiversity adversely. For the three sampling points, the BOD value ranged from 2.5- 3.5, 2.3-2.7, 6.1-7.9, and 5.5-6.4 mg/l in Nainital, Bhimtal, Naukuchiatal, Sattal, lakes respectively which is much better in comparison to the previous results and within the permissible limits (<6). Nainital and Bhimtal have a much better BOD level, the reason being much care has been shown for their restoration.

2. D.O.: A DO level of 5-12 mg/l has been considered desirable at the surface of the lakes. According to the CPCB, a water body which is used as a drinking water source must have D.O. of 6 mg/l, for bathing purpose it should be 5 mg/l and 4 mg/l for propagation of wildlife and fisheries. The D.O. level in the lakes is very good, especially in the Nainital lake, it has reached to a level of about 9 mg/l, the reason behind this is the aeration of the lake bed by artificial techniques. The Do level in rest of the lakes is also satisfactory.

3. Total Dissolved Solids: It is the term used to describe the inorganic salts and small amounts of organic matter present in solution in water. The principal constituents are usually calcium, sodium, magnesium and potassium cations and carbonate, chloride, hydrogencarbonate, sulfate, and nitrate anions. The desirable limit for TDS is below 500 mg/l. In this respect, all the four lakes have sufficient low amount of TDS.

4. pH: It is defined as the negative logarithm of hydrogen ion concentration in a solution. pH was found slightly alkaline in nature. The pH value of different samples were within prescribed range by WHO i.e. 6.5- 8.5.

5. Turbidity: It is a measure of the degree to which the water loses its transparency due to the presence of suspended particles. The lake water is slightly turbid. As compared to the previous studies, turbidity level has gone down slightly ranging from 7.6-8.25, 3.7-4.3, 5.9-6.5 and 7.25-8.1 in Nainital, Bhimtal, Naukuchiatal, Sattal lakes respectively. Higher turbidity level leads to increase in water temperature because suspended particles absorb more heat. Concentration of dissolved oxygen (DO) reduces in turn because warm water holds less DO than cold. The amount of light penetrating the water also reduces due to higher turbidity which reduces photosynthesis.

6. Copper: The desirable value for copper has been given as <5 mg/l and the level of copper ions is well below the permissible limits. It ranges from 0.013-1.09 in these lakes which is very much acceptable.

7. Iron: The desirable value for iron is set below 1.5 mg/l and it ranges from 0.011 to 0.67 mg/l in these lakes which is again well below the permissible limits.

8. Chromium: The value of chromium ranges from 0.06 to 0.24 mg/l in these lakes and as compared to the EPA limits for drinking water, the desirable limit has been given as 100 microgram/l and in this respect chromium level in the lakes is acceptable.

9. Nitrate: The permissible value of Nitrate has been given as below 0.30 mg/l (LDA, Nainital) and as compared to the previous studies and the standards, the level of nitrate is well acceptable.

10. Chemical Oxygen Demand : The lake water quality for is classified either as oligotrophic or eutrophic. In oligotrophic lakes, having very clear water, COD should be less than 1mg/l that is required for oligosaprobic species such as rainbow trout. In general, the COD of oligotrophic and eutrophic lakes containing oligosaprobic fish such as smelt, should be less than 3 mg/l. In eutrophic lakes containing carp, the COD should be less than 5 mg/l (Water Quality Standards for Fisheries, 1965). Less than 8 mg/l COD is desirable for waters used for swimming. High COD interferes with oxygen transfer to the soil, resulting death of rice plants. Experimental results show that a COD of less than 6 mg/l are desirable for agriculture use. In general 8 mg/l of COD is acceptable for most industrial uses and for conservation of environment . In this respect Cod level in the lake is acceptable.

CONCLUSION

The human activities shared with changes in land use prototype like agricultural, urbanization, deforestation etc. Greatly influence the water quality of the lakes. Urban settlements and increasing deforestation, pooled with rapidly increasing requirement for water, are causing more water quality management problems. The municipal wastes being biodegradable products predictable changes in water bodies and industrial effluent are responsible for pollution to a smaller extent but the effects created by them may be more severe as nature is often unable to understand them. Runoffs degrades the water quality of lakes as of there use from mineral fertilizers, chemical pesticides, which have mostly organic compounds. The main sources that pollutes these lakes are the continuous sedimentation rate, domestic effluents , agricultural runoffs through direct outfalls. In comparison to the Nainital lake, rest of the lakes are not given due importance by the administration.

REFERENCES

1. P.Z. Yanda , N.F. Madulu: Water resource management and biodiversity conservation in the Eastern Rift Valley Lakes, Northern Tanzania,((August 2005), Physics and Chemistry of the Earth 30 (2005) 717–725.
2. U.K. Shukla, D.S. Bora: Sedimentation model for the quaternary intramontane
3. Bhimtal–Naukuchiatal Lake deposits, Nainital, India (July 2002) : Journal of Asian Earth Sciences 25 (2005) 837–848
4. Preetam Choudhary, Joyanto Routh,, Govind J. Chakrapani: An environmental record of changes in sedimentary organicmatter from Lake Sattal in Kumaun Himalayas, India (April 2008): Department of Earth Sciences, Indian Institute of Technology Roorkee, Roorkee 247667, India,Department of Geology and Geochemistry, Stockholm University, 10691 Stockholm, Sweden : Science Of The Total Environment 407(2009)2783-2795.
5. Francesca Gherardi , J. Robert Britton , Kenneth M. Mavuti: A review of allodiversity in Lake Naivasha, Kenya: Developing conservation actions to protect East African lakes from the negative impacts of alien species (August 2010) : Biological Conservation 144 (2011) 2585–2596
6. G.A. Parisopoulos ,M. Malakou: Evaluation of lake level control using objective indicators: The case of Micro Prespa (November 2007) : Journal of Hydrology 367 (2009) 86–92
7. Marcus W. Beck, Bruce Vondracek: Environmental clustering of lakes to evaluate performanceof a macrophyte index of biotic integrity (November 2011): Aquatic Botany 108 (2013) 16– 25
8. Andreas Efstratiadis , Kimon Hadjibiros: Can an environment-friendly management policy improvethe overall performance of an artificial lake? Analysis of a multipurpose dam in Greece (June 2011): environmental science and policy 14(2011)1151-1162.
9. Francesca Gherardi , J. Robert Britton , Kenneth M. Mavuti: A review of allodiversity in Lake Naivasha, Kenya: Developing conservation actions to protect East African lakes from the negative impacts of alien species (August 2010) : Biological Conservation 144 (2011) 2585–2596